Structural Collapse at Dwelling Fire
Results in Two Firefighter Fatalities
Stockton, California

Investigated by: Dennis C. Duckett

This is Report 102 of the Major Fires Investigation Project conducted by Varley-Campbell and Associates, Inc./TriData Corporation under contract EMW-94-4423 to the United States Fire Administration, Federal Emergency Management Agency.

Homeland
Security

Department of Homeland Security
United States Fire Administration
National Fire Data Center

U.S. Fire Administration Fire Investigations Program

The United States Fire Administration develops reports on selected major fires throughout the country. The fires usually involve multiple deaths or a large loss of property. But the primary criterion for deciding to do a report is whether it will result in significant "lessons learned." In some cases these lessons bring to light new knowledge about fire--the effect of building construction or contents, human behavior in fire, etc. In other cases, the lessons are not new but are serious enough to highlight once again, with yet another fire tragedy report. In some cases, special reports are developed to discuss events, drills, or new technologies which are of interest to the fire service.

The reports are sent to fire magazines and are distributed at National and Regional fire meetings. The International Association of Fire Chiefs assists USFA in disseminating the findings throughout the fire service. On a continuing basis the reports are available on request from USFA; announcements of their availability are published widely in fire journals and newsletters.

This body of work provides detailed information on the nature of the fire problem for policymakers who must decide on allocations of resources between fire and other pressing problems, and within the fire service to improve codes and code enforcement, training, public fire education, building technology, and other related areas.

The Fire Administration, which has no regulatory authority, sends an experienced fire investigator into a community after a major incident only after having conferred with the local fire authorities to insure that USFA's assistance and presence would be supportive and would in no way interfere with any review of the incident they are themselves conducting. The intent is not to arrive during the event or even immediately after, but rather after the dust settles, so that a complete and objective review of all the important aspects of the incident can be made. Local authorities review USFA's report while it is in draft. The USFA investigator or team is available to local authorities should they wish to request technical assistance for their own investigation.

This report and its recommendations were developed by USFA staff and by Varley-Campbell & Associates, Inc. Miami and Chicago, its staff and consultants, who are under contract to assist the Fire Administration in carrying out the Fire Reports Program.

The United States Fire Administration greatly appreciates the cooperation received from the Stockton Fire Department in preparing this report.

For additional copies of this report write to the United States Fire Administration, 16825 South Seton Avenue, Emmitsburg, Maryland 21727. The Report and the photographs in color are available on the Administration's Web page at http://www.usfa.dhs.gov/

U.S. Fire Administration

Mission Statement

As an entity of the Department of Homeland Security, the mission of the USFA is to reduce life and economic losses due to fire and related emergencies, through leadership, advocacy, coordination, and support. We serve the Nation independently, in coordination with other Federal agencies, and in partnership with fire protection and emergency service communities. With a commitment to excellence, we provide public education, training, technology, and data initiatives.

TABLE OF CONTENTS

Stockton, California
Structural Collapse At Dwelling Fire
Results In Two Firefighter Fatalities
February 6, 1997

Investigated by: Dennis C. Duckett

Written by: Sheila-Faith Barry
J. Gordon Routley

Local Contacts: Douglas C. Ratto, *Fire Chief*
Gary Gillis, *Division Chief*
Marty Galindo, *Inspector*

Stockton Fire Department
110 West Sonora Street
Stockton, California 95203

OVERVIEW

Two firefighter fatalities and one civilian fatality occurred in a single family residential dwelling fire. The second floor addition collapsed during suppression operations, trapping the firefighters. A captain was also trapped under the debris, sustaining serious injuries, but was later successfully rescued.

On Thursday, February 6, 1997, The Stockton Fire Department responded to twelve 9-1-1 calls reporting a large fire. The first firefighters on the scene at 4:16 a.m. found what appeared to be a one story home in flames. Because of the small lot size, a perimeter fence and a crowded driveway, there was very limited access to the sides and rear of the home. The constraining perimeter and heavy volume of flames at the rear of the home made it difficult to view the back of the house.

The elderly female resident was reported to still be inside the home, so an interior attack and search for the victim ensued. Firefighters moving towards the rear of the home were unaware of the existence of a second story addition, which was not visible from the front of the house. The addition, which was inadequately supported by its first floor construction, subsequently collapsed, pinning two firefighters and a captain under heavy rubble.

Efforts to extricate the trapped firefighters were immediately initiated. Cribbing and hydraulic rescue tools were used to lift and support the heavy debris; firefighters were able to locate and rescue the fire captain. The two firefighters were subsequently removed from the debris. The body of the occupant was later located in the area of the collapsed second floor master bedroom.

SUMMARY OF KEY ISSUES

ISSUES	COMMENTS
Fire origin and cause	The fire was determined to have started accidentally in the master bedroom on the second floor.
Firefighter Fatalities	Two firefighters were trapped and killed and a captain was seriously injured when the second floor collapsed.
Structure	The section that collapsed was a two-story addition at the rear of an originally single-story dwelling. The addition was not visible from the front and crews working inside did not realize there was a floor above them. The addition had a large open area on the lower level and its construction inadequately supported the second story.
Rescue	Rescue efforts were hampered by difficulty accounting for and locating the trapped firefighters. Cribbing and hydraulic rescue tools were needed to raise the heavy debris.
Pass Devices	Pass devices (although worn) weren't activated and did not contribute to locating the trapped firefighters.
Protective Equipment	The trapped firefighters and captain were wearing full protective equipment, including pass devices and SCBA.

FIRE DEPARTMENT

The Stockton Fire Department protects approximately 235,000 residents in a city of 55.7 square miles. The department has twelve stations and is staffed by 231 career uniformed members and 27 civilian personnel. The department provides both fire suppression and emergency medical response. A private contractor operates the ambulance service.

The Fire and EMS Communications Center, which dispatches the Stockton Fire Department, four surrounding fire districts and the ambulance service, is staffed by a supervisor and three telecommunicators on each shift. Stockton has mutual aid agreements with the four surrounding fire protection districts. The Stockton Auxiliary Firefighters respond to second and greater alarms, however the auxiliary firefighters do not participate in interior attack.

Stockton's twelve engine and three truck companies normally operate with 4-person crews, with the exception of Truck 2, which operates with 5 crew members. Engine companies are staffed by a captain, an engineer and two firefighters, while truck companies are staffed by a captain, an engineer, a tiller operator and one firefighter. Support Unit 2 is staffed by a captain and a firefighter/paramedic. Two battalion chiefs are on duty at all times and a chief's operator responds in a separate vehicle to perform command post support functions.

THE HOUSE

The house at 26 West Mendocino Avenue was originally built in 1942 as a 5-room, 1-1/2 story, single family dwelling. The 900 square foot floor plan was typical of houses in the neighborhood of single family dwellings on narrow residential streets, about one mile north of the downtown business district. It was built on a 52-foot by 100-foot lot in the middle of a residential block.

The interior and roof construction of the house were wood frame, however the exterior walls that were visible from the front were brick. A 2-story wood frame section, with wood exterior walls, was added to the rear of the house in 1952. The addition, which could not be seen from the street, almost doubled the ground floor area of the house and included a large master bedroom suite and a sundeck on the upper level. The lower level of the addition was a large open room, approximately 37 feet by 27 feet in area, which was used by the occupant as a dance studio.

To create this large open area on the lower area, the upper floor was constructed with 3 by 12 wood joists on 12-inch centers supporting a hardwood floor. The joists, which were almost 22 feet long, were supported at one end by an exterior bearing wall and by a header beam, which ran across the room at an unusual angle. Part of the upper story and the sundeck were cantilevered beyond the header beam.

The header beam was constructed of doubled 2 by 12 wood members assembled with a narrow space between them. It was attached at an irregular angle to the rear wall of the original structure and to the rear exterior wall of the addition. A single 4 by 6 wood post supported the header beam within the room.

The dance studio was constructed with a hardwood floor and with 1 by 8 tongue and groove wood covering on most of the walls and ceiling. The exterior walls had several windows and sliding glass doors. On the upper level, sliding glass doors opened to the sundeck, which ran along the east side of the addition. The only access to the upper level was a narrow stairway that rose from the studio along the west wall of the addition near the original rear wall to the master bedroom.

The lot was crowded with the enlarged dwelling and a 500 square foot wood garage at the rear of the property, almost touching the house. Six-foot high fences were built on the property lines at the rear and along most of both sides, leaving a narrow space along the west side of the property as the only access to the small rear yard. At the time of the fire, three automobiles and a van were parked in the driveway that ran along the east side, further congesting access to the rear of the property.

The only resident at the time of the fire was an 82 year old woman, who was apparently sleeping in the master bedroom on the second floor. She was a retired dance instructor and had lived in the house for many years. The fire was determined to have originated in the bedroom area, possibly from an electric blanket, and extended down through the floor into the lower level. The occupant was asphyxiated by carbon monoxide, which suggests that the fire smoldered for a considerable time before the rapid fire spread began.

THE FIRE

The first call to 9-1-1 came in at 4:11 a.m., reporting a "big" fire at 50 West Mendocino Avenue. This was followed immediately by a second caller reporting a fire in a neighbor's garage. Several additional calls followed in rapid succession reporting a serious fire, but giving different addresses in same neighborhood. A total of twelve calls to 9-1-1 were received, including one from a security guard at the nearby University of the Pacific campus who reported hearing popping noises and explosions.

FIRE DEPARTMENT RESPONDS

The initial response of Engine 9, Engine 6, Truck 4, Support 2, Battalion 2 and Operator I was dispatched at 4:12 a.m. to a reported garage fire at 50 West Mendocino Avenue, near the intersection of Commerce. An additional company, Engine 4 was dispatched to respond "code 2" (non-emergency) as the rapid intervention team (RIT).

Arriving on scene at 4:16 a.m., the Captain of Engine 9 observed and reported heavy fire showing from the rear and through the roof of the house at 26 West Mendocino. Two parked cars were burning in the driveway on the east side of the house and the exterior siding and the roof overhang of the adjoining house at 18 West Mendocino had been ignited by radiant heat. Neighbors on the street informed the E9 Captain that the elderly female resident of the burning house was believed to still be inside.

Engine 9 requested a second alarm at 4:16 a.m., reporting two structures involved, with a possible occupant trapped in one of the houses. The captain directed the two firefighters from E9 to force open the front door and advance a 1-3/4 inch preconnected attack line into the home.

Arriving on scene at 4:18 a.m., Engine 6 laid 400 feet of 5-inch supply line from a hydrant on Center Street to provide a supply line for Engine 9. While the supply line was being connected, the E6 Captain directed one of his crew members to take the booster line from Engine 9 to cover the exposures on the east side. The E6 Captain then initiated a rapid search of the exposed house at 18 West Mendocino, along with Truck 4's captain and firefighter.

Battalion Chief 2 arrived at 4:19 a.m. and assumed command of the incident, while a command post was established by Operator 1 at the corner of Mendocino and Center Streets, approximately 200 feet east of the fire. The battalion chief attempted to conduct a perimeter survey of the fire, however the volume of fire and the fences restricted his access. He did note that the rear of the burning structure had two stories and the fire had extended to the garage at the rear of the lot.

Support Unit 2 arrived at the same time and immediately called for the electric company to respond, as energized wires were down and arcing. The captain advised the Incident Commander that he and a firefighter would advance a 2-1/2 inch line from Engine 9 to attempt to knock down the large volume of fire on the east side where the cars, garage and fence were burning.

Engine 4, which had upgraded to "Code 3" (emergency) response, arrived at 4:20 a.m. and was directed by the Incident Commander to cover the exposure on the west side (at 36 West Mendocino). This crew took the second 1-3/4 inch preconnected line from Engine 9 and worked their way down the west side. They controlled flames that had ignited the exterior of the adjoining house.

INTERIOR ATTACK

The E9 Captain and two firefighters entered 26 West Mendocino through the front door with the 1-3/4 inch hoseline. Encountering moderate heat and very heavy smoke conditions, with no visible flames, they proceeded through the older section toward the rear of the house. The E9 Captain made an inspection hole in the ceiling with a pike pole, but found no fire in the attic area of the original structure.

The E6 Captain and one of the firefighters then deployed a standpipe pack to advance a back-up line into the interior. While the other firefighter fed the line from the door, the Captain advanced the 1-3/4 inch line toward the rear of the house. Noting that Engine 9's line had gone to the right, he took the back-up line toward the left side of the house, where fire was visible in the kitchen area. (See Figure 2)

The Truck 4 Captain, engineer and firefighter also entered through the front door at that time. While the engineer assisted the E6 Captain, who was advancing the back-up line, the T4 Captain and firefighter attempted to perform an interior search. Their search effort was limited by the heat and smoke conditions, so the T4 Captain went back outside to look for an alternative access to the rear, while the firefighter continued the search. Finding no alternative access, he returned to the interior to continue the search.

While the T4 Captain was outside, the firefighter from Truck 4 had encountered the crew on the back-up line in the kitchen. The E6 Firefighter, who had been feeding the line at the door, had entered and followed E9's line until he joined with the E9 crew. At this point, both lines were operating on the fire in the large room at the rear of the ground floor, but none of the personnel inside the house were aware of the second floor above that area.

SECOND ALARM COMPANIES

The second alarm assignment of Engine 11, Engine 2, Truck 2, Medic 72 and Battalion 1 had been dispatched at 4:17 a.m. All Stockton Fire Department administrative staff officers were also alerted on the second alarm.

Engine 2 arrived at 4:22 a.m., followed by Truck 2 and Battalion 1 at 4:23 a.m. While Truck 2 proceeded to ladder and check for extension to the houses east and west of the initial fire, the E2 Captain made his way to the rear yard of the east exposure. From this location he observed that the rear of the burning house was fully involved in flames, however there was so much fire at the rear of the house that he did not identify that there was a second floor.

Engine 2 and Battalion 1 were directed to evaluate conditions on the south side of the fire. They positioned Engine 2 in front of 25 West Mariposa Avenue and determined that they would need hoselines to protect the exposures bordering the rear of the involved structure. There was no hydrant on the block and they requested another engine company from Incident Command to lay a supply line to Engine 2. Engine 11, which had been directed to approach from the west, because West Mendocino Avenue was heavily congested east of the fire, was redirected to provide the water supply for Engine 2. E4's engineer was also directed to take the apparatus to the south side and assist with the water supply.

The crew of Engine 2 advanced a 2-1/2 inch line and a 1-3/4 inch line to the rear yard of 25 West Mariposa, to the rear of the structure, and directed a stream onto the garage and the rear of the burning house, as well as a burning power pole. A concrete block fence and arcing power lines prevented their access to the back yard of 26 West Mendocino. Engine 11 advanced a second 2-1/2 inch line to the same area, but the line was not put into operation.

By 4:29 a.m., Battalion 1 had advised the Incident Commander (Battalion Chief 2) that the 2-1/2 inch line at the rear could hit the fire on the second floor through a window. Support 2 also advised that they could hit the fire on the second floor with their 2-1/2 inch line from their position on the east side. The Incident Commander approved these actions, warning the exterior crews that there were crews operating in the front part of the house. At 4:31 a.m., Battalion 1 advised the command that they were getting the fire in the rear of the house knocked down and the Incident Commander noted that the fire appeared to be almost under control.

Operator 1, at the command post, reminded the Incident Commander that the "safety engine" had been committed and there was no rapid intervention team assigned. At 4:32 a.m., Battalion 1 released Engine 11 to report to the command post as a relief company. The engineers from Engine 6 and Engine 11 were ordered to report to the front of the house as a rapid intervention team, while the remainder of Engine 11's crew prepared to enter and relieve one of the interior crews. The Incident Commander directed Engine 11 to relieve the crew from Engine 9, who had been operating inside the house for almost 15 minutes.

At 4:34 a.m., Battalion 1 informed the Incident Commander that flames were again beginning to build up in the rear of the building. The hose stream was again directed into the second floor window, darkening down the flames. Two minutes later, Battalion 1 reported that there was still an active fire on the second floor

CREW ROTATION

When their SCBA low air warning bells began to sound, the E9 Captain and one firefighter from his unit exited the house, leaving two firefighters, one from Engine 9 and one from Engine 6 on the attack line. After exchanging their air cylinders, they reentered and relieved the other E9 firefighter, who then exited to exchange his air cylinder. The E6 Captain also exited to exchange his air cylinder, leaving two crew members from Truck 4 on the back-up line, while the other two crew members from Truck 4 were conducting a search in the section of the original structure at the opening of the addition on the ground floor. At this point most of the visible flames in the ground floor interior had been knocked down and the attack line had been advanced into the large room. The back-up line was at the doorway between the kitchen and the rear part of the house.

The E9 Captain then moved toward the east to investigate a glow, leaving the two firefighters with the attack line. He followed the back-up line into the kitchen, where he met up with the two crew members from Truck 4. The E9 Captain and the other Truck 4 crew member, who had completed their search, were moving toward the front of the house to exit, when they passed the E6 Captain, who had reentered and was moving toward the rear.

SECOND FLOOR COLLAPSES AND SUBSEQUENT RESCUE

At 4:37 a.m., as the E6 Captain was stepping over the threshold from the kitchen into the addition, the second floor addition collapsed with a loud crack and rumble. The E6 Captain was trapped under the debris, along with the E9 firefighter and the E6 firefighter who were advancing the attack line, several feet inside the addition at the rear of the house. The two T4 crew members who were operating the back-up line were pushed back into the kitchen area by the falling debris and narrowly escaped being trapped.

Battalion 1 observed the collapse from his position and immediately radioed to Incident Command to request a personnel accountability check. Engine 4 and Support 2, operating outside the structure, also both reported a structural collapse at the rear. T4's Captain stepped just outside the front door and radioed Command, advising of a wall and ceiling collapse with firefighters trapped. Less than a minute later, he reported that two firefighters were trapped and that airbags might be needed to rescue them.

E11's Captain then reported that four or five firefighters were working to free trapped personnel, who were pinned under a 20 foot wall section. He estimated that the debris weighed several hundred pounds. He also requested a back-up line through the front door. The captains of Truck 2, Truck 4 and Engine 11 began to direct rescue operations inside the structure.

Battalion 1 reported that they had a back-up line coming in from the rear. The crew of Engine 2 extended their 2-1/2 inch line with 1-3/4 inch hose and advanced the line over the perimeter wall to knock down the fire in the collapsed section. No flames were observed around the trapped fire-fighter, but the area was wet down to prevent ignition.

Three chief officers, who had responded from their homes on the second alarm, had just reported to the command post when the collapse occurred. The deputy chief assumed command of the incident, reassigning Battalion 2 as the Operations Chief. A division chief was assigned as Safety Officer and a battalion chief was assigned as the Rescue Group Supervisor. A staff firefighter, who had also responded on the second alarm, was assigned to perform a face-to-face accountability check of the units on the scene.

RESCUE EFFORTS BEGIN IMMEDIATELY

The rescuers could see one trapped individual under the collapsed second floor and believed it was the E9 firefighter. Using a hydraulic rescue tool and cribbing, the rescuers were able to lift the collapsed floor enough to rescue this individual and discovered that it was the E9 Captain. He was removed at 4:48 a.m., about 10 minutes after the collapse, and transported to the hospital in serious condition, accompanied by the firefighter/paramedic from Support Unit 2.

Another trapped firefighter was located about 11 feet further into the collapsed area. Using a second hydraulic rescue tool and more cribbing, the rescuers were able to create a space of approximately 14 inches under the debris in which to maneuver. Two firefighters crawled under the raised debris and were able to reach the trapped firefighter. They discovered that this victim was the E6 firefighter, who was in respiratory arrest. However, given his confined space location, there was not enough space to perform artificial respiration and additional effort would be required to extricate him.

At 5:11 a.m., Engines 1 and 5 were requested as relief companies. Three minutes later Truck 3 was called to respond with additional cribbing and rescue equipment. The rescue operations were conducted using hydraulic rescue tools, a floor jack and cribbing to lift the heavy collapsed floor sections. Although air bags were available at the scene (carried on Support Unit 2), their use was ruled out because of nails in the debris and the smoldering fire

Looking deeper into the debris, the rescuers spotted the reflective material on the turnouts of another firefighter. At 5:28 a.m., the Rescue Group Supervisor reported that two firefighters were still trapped in the debris. The firefighter from Engine 6 was removed almost an hour after the collapse at 5:44 a.m. and transported to the hospital, accompanied by the same firefighter/paramedic who had just returned from the hospital.

The third victim was removed at 6:19 a.m. and confirmed to be the E9 firefighter. At this point, the Rescue Group Supervisor advised the Incident Commander that rescue operations had been completed.

Crews leaving the scene were directed to Station 2 to participate in Critical Incident Stress Debriefing, while Engines 3 and 14 were dispatched to the scene to search for the occupant of the home and to continue overhaul. A K-9 team located the body of the occupant in the debris of the collapsed second floor master bedroom, later in the day.

INJURIES AND FATALITIES

Both firefighters were pronounced dead at the hospital from asphyxiation and crushing injuries. They both appeared to have died as a result of the weight of the structural debris that fell on them.

The E9 firefighter had been with the Stockton Fire Department for only three months and was still a probationary firefighter - this incident was his first working structure fire. The E6 firefighter had previously been employed as a firefighter by the California Department of Forestry and had been a member of the Stockton Fire Department since 1990.

The E6 Captain suffered second and third degree burns and a broken rib.

ANALYSIS

Structural Analysis

Analysis of the structure indicated that the heavy second floor was inadequately supported and vulnerable to the type of catastrophic collapse that occurred. The direct cause of the collapse was attributed to either a failure of the 4 by 6 support post within the room or the failure of the header beam itself.

It would have been difficult to anticipate the collapse that occurred without prior knowledge of the structure or a detailed size-up of the structure at the time of the fire. Large open ground floor areas with extensive wall openings are normally considered to have collapse potential. However, the potential for collapse in the building was difficult to ascertain due to the large volume of flames and the lack of accessibility to view the structural components.

Several firefighter fatalities have occurred in structures that presented significantly different appearances from different viewing positions, including several instances where interior crews were unaware of other floor levels either above or below them. The presence of differing construction (i.e., wood frame addition to a brick structure) should also be an indicator of potential structural problems. Observations of this nature should be clearly communicated, to ensure that interior crews will be aware of structural features identified by exterior personnel. The arrangement of the structure with a two-story addition at the rear of what appeared to be a single story dwelling, should have been a warning.

Accountability

The lack of crew integrity and accountability was an important factor at this incident. Instead of working as cohesive teams, crew members were separated and rearranged themselves to perform different tasks and functions as the circumstances unfolded. Company officers were often unaware of the location and function of their respective crew members and several personnel working inside the structure were unsupervised and not in radio contact with the Incident Commander at different times. The lack of communication became critical when the collapse occurred and no one knew who was in the area or who was missing.

After the collapse occurred, several attempts were made to account for all of the personnel on the scene and determine who was missing. The first request for an accountability check came from Battalion 1, seconds after the collapse occurred, however a full accounting for personnel was not accomplished until the last victim had been removed, more than an hour and a half later. Two of the three victims who were removed had not been reported as missing.

The first attempt at an accountability check was assigned to a firefighter a few minutes after the collapse occurred. He attempted a face-to-face check of all crews on the scene, however this was unsuccessful due to the confusion and constant movement of personnel as they attempted to rescue the trapped victims.

At 4:55 a.m. an off-duty chief's operator arrived at the command post and took over the accountability process, attempting a roll-call of the units by radio. This was also unsuccessful due to the number of interruptions caused by urgent messages on the tactical radio channel and the difficulty of contacting company officers who were engaged in the unfolding rescue operation.

Incident Management

The first arriving company officer (E9 Captain) gave a report on visible fire conditions and called for a second alarm, but did not provide any further direction for incoming units. Incident command was established when the battalion chief arrived. However, most of the tactical functions performed by individual companies appeared to have been self-initiated and uncoordinated. The existence and condition of the second floor addition was not initially recognized or clearly communicated to the interior crews. While the volume of the fire was significant, it did not initially appear to be unusually dangerous or challenging and there was no recognition that the incident involved unusual structural hazards.

During the initial attack stage, the Incident Commander (Bat 2) was located near the front of the fire building, while a remote command post was staffed by the chief's operator. Most of the radio traffic, including several assignments to incoming units, was handled by the operator, who was not physically located with the Incident Commander. The operator also relayed messages between Battalion Chief 1 who was outside at the rear and the Incident Commander (Bat 2) at the front of the building. The use of the operator to relay information appears to have contributed to the overall confusion and a lack of coordination among units, particularly when the units at the rear were reporting fire conditions that could not be observed by the Incident Commander at the front of the building.

The Incident Commander attempted to size-up the situation from the exterior and his major concern appeared to be exterior exposures, although there was an interior attack in progress with a report of a trapped occupant. The extra company that was dispatched to provide a rapid intervention team was reassigned to cover an exposure and the RIT assignment was not covered for several minutes. No safety officer was assigned until after the collapse occurred.

The Incident Commander did not receive reports on interior conditions from the crews working inside the house and apparently did not provide direction to them. As the threat of extension to the exposure was controlled, the authorization was given to hit the fire on the second floor through the windows, however the interior crews were not specifically warned. While this action did not appear to have caused the collapse, the notification would have made the interior crews aware of the upper floor and the associated risk. The radio transcript suggests that the Incident Commander did not realize that the interior crews had worked their way into the rear portion of the structure.

The incident management structure expanded when the off-duty command officers arrived at the command post as a result of the second alarm. The absence of support units that would normally deploy to an incident such as this and sufficient radios to ensure that all officers were able to communicate minimized the effectiveness of these additional resources.

Protective Clothing and Equipment

The protective clothing, SCBA's, PASS devices and radios assigned to the trapped firefighters and captain were tested and determined to be functional. The deaths and injuries do not appear to be related to any failure to use protective equipment or any shortcomings of the equipment itself. However, there were no reports that PASS devices could be heard or the alarming units were helpful in locating the trapped individuals, suggesting that their PASS devices may not been activated.

LESSONS LEARNED

1. **Unusual details of the arrangement of burning structures must be communicated to interior crews.**

   Having approached the burning structure from the front (north) side, crews working inside the house thought they were fighting a fire in a single story dwelling. From their vantage point in the rear of the structure, the Battalion Chief and the crew of Engine I observed a second floor; these observations were not clearly communicated to the interior crews. All information regarding a structure, even information which may appear obvious, should be reported to Incident Command to ensure that the information is passed along to company officers and interior crews.

2. **Large open areas and non-standard construction are indicators of potential structural collapse.**

   The heavy floor structure that was used to create a large open area on the ground floor of the addition, which was inadequately supported, was vulnerable to sudden structural failure. Fire department personnel should be aware that unusual structural components can often contribute to unexpected collapse.

3. **The Incident Commander must direct the operation.**

   Situations where company operations are self-initiated and self-directed often lack coordination and control. Uncoordinated actions by firefighting crews generally detract from an efficient and safe operation. The Incident Commander is responsible for the overall operation, and therefore should establish and maintain command throughout the operation. Although specific functions and activities can and often should be delegated, the Incident Commander should receive updates and be informed of any new information immediately, so that information can be acted upon.

4. **The Incident Commander must control the communications process.**

   The assignment of an operator to handle radio traffic can provide valuable assistance to an Incident Commander, however the Incident Commander must be aware of all important messages and direct all assignments. The ability of an Incident Commander to effectively command fire operations is directly related to the timeliness and clarity of the information available to him. If used, an operator should be located in the command area with the Incident Commander to ensure ease of information transfer.

5. A standard accountability system should be utilized at every incident.

There was no formal personnel accountability system in place prior to the collapse. The establishment and maintenance of a personnel accountability system is a critical component in all tactical operations. These unsuccessful efforts point out the necessity to establish an accountability system before something goes seriously wrong at an incident scene. Attempting to initiate an accountability process after the situation has become critical is likely to have similar limited results. When something does happen, the system should be in place and the crews should have the unit discipline and self-control to quickly account for all personnel and report their status to the Incident Commander. The establishment of a personnel accountability system is a pre-requisite to a safe and effective operation.

6. Attempts to rescue trapped or missing firefighters require a high level of coordination, control and risk management.

The urgency of the mission demands that these operations be conducted as quickly as possible, but without proper planning and organization, risk to rescuing firefighters may result. Rapid intervention teams (RIT), outfitted with appropriate equipment, provide an immediate resource and should be provided. Demands for additional firefighting resources should not be cause to disband, even temporarily, a RIT.

7. Effective communication is dependent upon the availability of communication equipment.

The Stockton Fire Department's mobile command vehicle was out of service. Without this vehicle, the number of extra radios available at the scene was severely limited. The additional staff officers who arrived in response to the second alarm were unable to be fully utilized because of the lack of radios, suggesting that there should be an alternative vehicle that could deliver essential equipment to the scene of a major incident.

8. Smoke detectors provide early warning to occupants of a fire.

Reports indicated that the elderly occupant of the home died from smoke inhalation, most likely from a smoldering fire condition that persisted long before the first flames were viewed by neighbors. A working smoke detector may have alerted the occupant of the fire, saving her life.

9. Second exits provide a second chance for escape.

There was only one exit from the bedroom area located on the second floor of the addition. This exit was an interior stairway to the first floor. A second exit provides an escape option in a fire if the main exit is blocked by flames or heavy smoke.

APPENDIX A

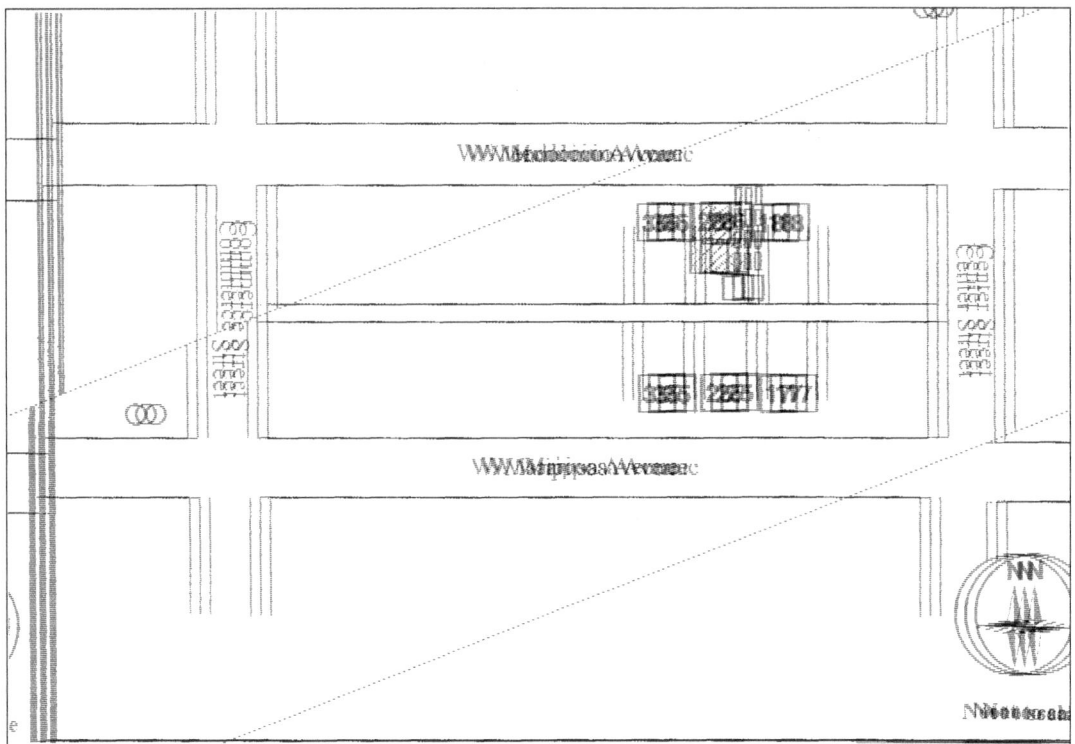

Figure 1: Site Plan

Figure 2: Floor Plan – First Floor

Figure 3: Floor Plan – Second Floor

APPENDIX B

Photographs Were Obtained From the Stockton Fire Department

1. View of #25 West Meadows from the street (north) side

2. Looking south towards rear of house -- note area of second floor collapse

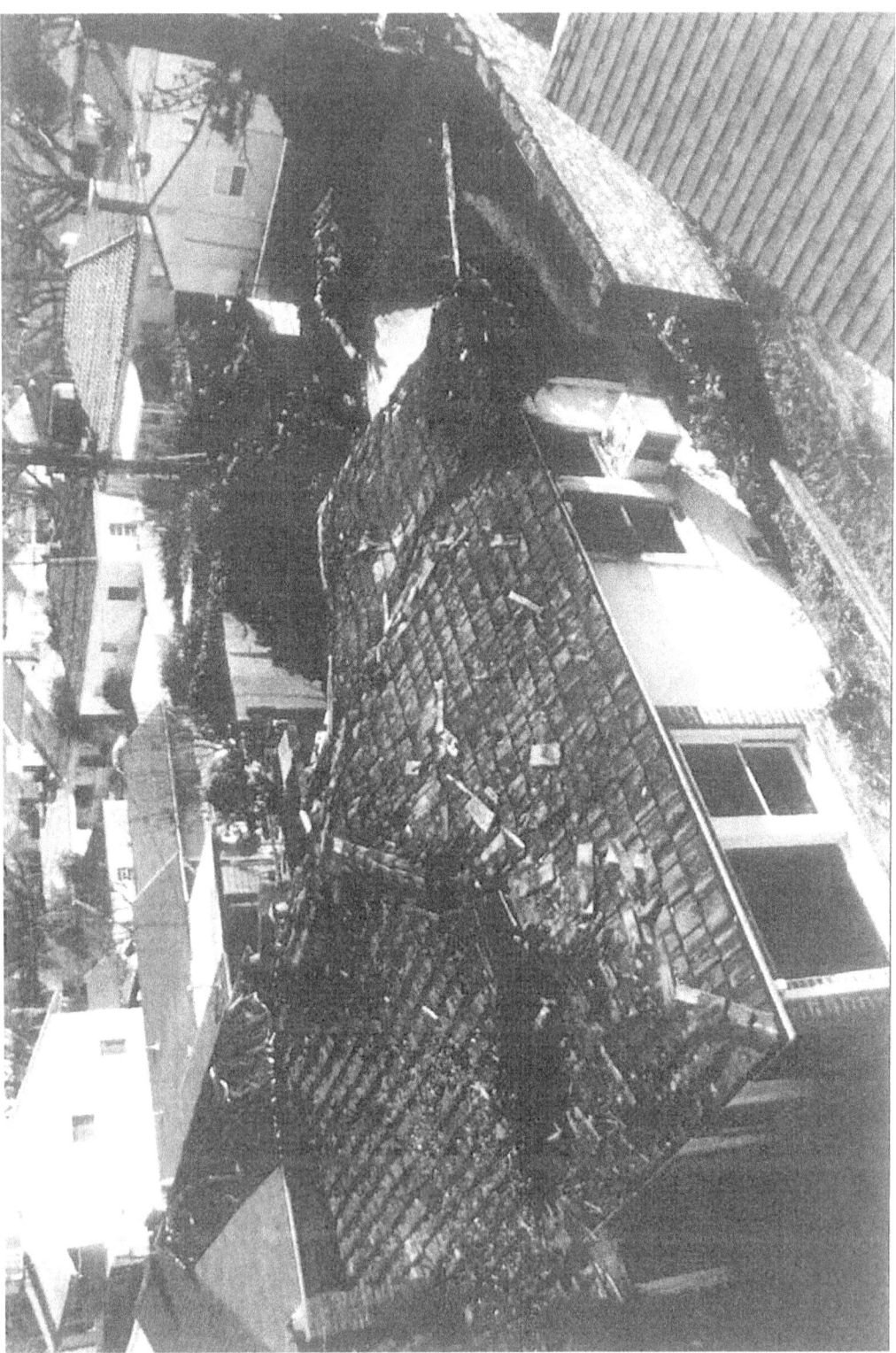

3. Southwest side of house

4. Looking southwest, across back room addition, from the driveway. The original house is on the right side of the photograph.

www.ingramcontent.com/pod-product-compliance
Lightning Source LLC
Chambersburg PA
CBHW081248170526
45165CB00009B/3240